DODD, MEAD WONDERS BOOKS include WONDERS OF:

BADGERS. Lavine
CATTLE. Scuro
CORALS AND CORAL REEFS.
    Jacobson and Franz
COYOTES. Lavine
CROWS. Blassingame
DONKEYS. Lavine and Scuro
DRAFT HORSES. Lavine and
    Casey
DUST. McFall
EAGLE WORLD. Lavine
EGRETS, BITTERNS, AND
    HERONS. Blassingame
ELEPHANTS. Lavine and Scuro
FLIGHTLESS BIRDS. Lavine
FROGS AND TOADS. Blassingame
GEESE AND SWANS. Fegely
GOATS. Lavine and Scuro
HIPPOS. Lavine
LIONS. Schaller
MARSUPIALS. Lavine
MICE. Lavine

MULES. Lavine and Scuro
PEACOCKS. Lavine
PIGS. Lavine and Scuro
PONIES. Lavine and Casey
RACCOONS. Blassingame
RATTLESNAKES. Chace
RHINOS. Lavine
SEA HORSES. Brown
SEALS AND SEA LIONS. Brown
SHARKS. Blassingame
SHEEP. Lavine and Scuro
SNAILS AND SLUGS. Jacobson and
    Franz
SPONGES. Jacobson and Pang
TURKEYS. Lavine and Scuro
TURTLE WORLD. Blassingame
WILD DUCKS. Fegely
WOODCHUCKS. Lavine
WORLD OF BEARS. Bailey
WORLD OF HORSES. Lavine and
    Casey
ZEBRAS. Scuro

Wonders
of Badgers

Sigmund A. Lavine
ILLUSTRATED WITH PHOTOGRAPHS
DRAWINGS, AND OLD PRINTS

DODD, MEAD & COMPANY
New York

*For Rosemary
who has badgered me for years*

ILLUSTRATIONS COURTESY OF: Author's collection, 6, 8, 22, 23, 33, 43, 50, 52, 56; Field Museum of Natural History, Chicago, 27, 54; Forestry Commission of Great Britain, 25, 38, 47, 60; Nicholas J. Krach, 3, 13, 20, 28, 35, 55; Jane O'Regan, 30, 31, 41; Leonard Lee Rue III, 29, 32, 40, 44; USDA, 2, 58; Wisconsin Division of Tourism, 49.

FRONTISPIECE: *Photograph of an American badger*

TITLE PAGE: *Old print of a European badger*

Copyright © 1985 by Sigmund A. Lavine
All rights reserved
No part of this book may be reproduced in any form
without permission in writing from the publisher
Distributed in Canada by McClelland and Stewart, Limited, Toronto
Manufactured in the United States of America

1 2 3 4 5 6 7 8 9 10

Library of Congress Cataloging in Publication Data

Lavine, Sigmund A.
  Wonders of badgers.

  Includes index.
  Summary: Discusses the characteristics of American and European badgers, as well as more exotic species, and includes lore that has grown up around these intelligent and often misunderstood creatures.
  1. Badgers—Juvenile literature. [1. Badgers]
I. Title.
QL737.C25L384 1985   599.74'447   84-25941
ISBN 0-396-08581-4

Contents

1. Meet the Badger 7
2. Lore of the Badger 12
3. Physical Characteristics 24
4. Badger Behavior 37
5. Exotic Badgers 51
6. Badger and Man 57
   Index 62

An excellent 1707 representation of a badger, a glutton (wolverine), and a zoril, all Mustelidae

"Man has a natural desire to learn."
—*Butler*

1 Meet the Badger

APPROXIMATELY sixty million years ago, a group of small, active animals with well-developed brains moved out of their original home in the trees and became ground-dwellers. These creatures, known as miacids (the name means they had cutting teeth) were meat-eaters. They came down to the Earth's surface in hopes of finding new sources of food.

More intelligent than the animals on which they preyed, the miacids prospered. So did their descendants, which eventually developed into numerous new species. However, irrespective of their size, habits, or habitat, all these animals retained the strong jaws and sharp teeth of the miacids. They needed both—like their ancient ancestors, the miacids' heirs were meat-eaters.

Zoologists (students of animal life) call meat-eating animals, including man, carnivores. This name was derived from the Latin *caro* (flesh) and *vovare* (to devour). Scientists have organized the four-legged meat-eaters into three main groups. Two of them are obvious—the cat tribe and the dog tribe. The third category is not as evident to the layman. It consists of the seals, aquatic carnivores that feed on the flesh of fish.

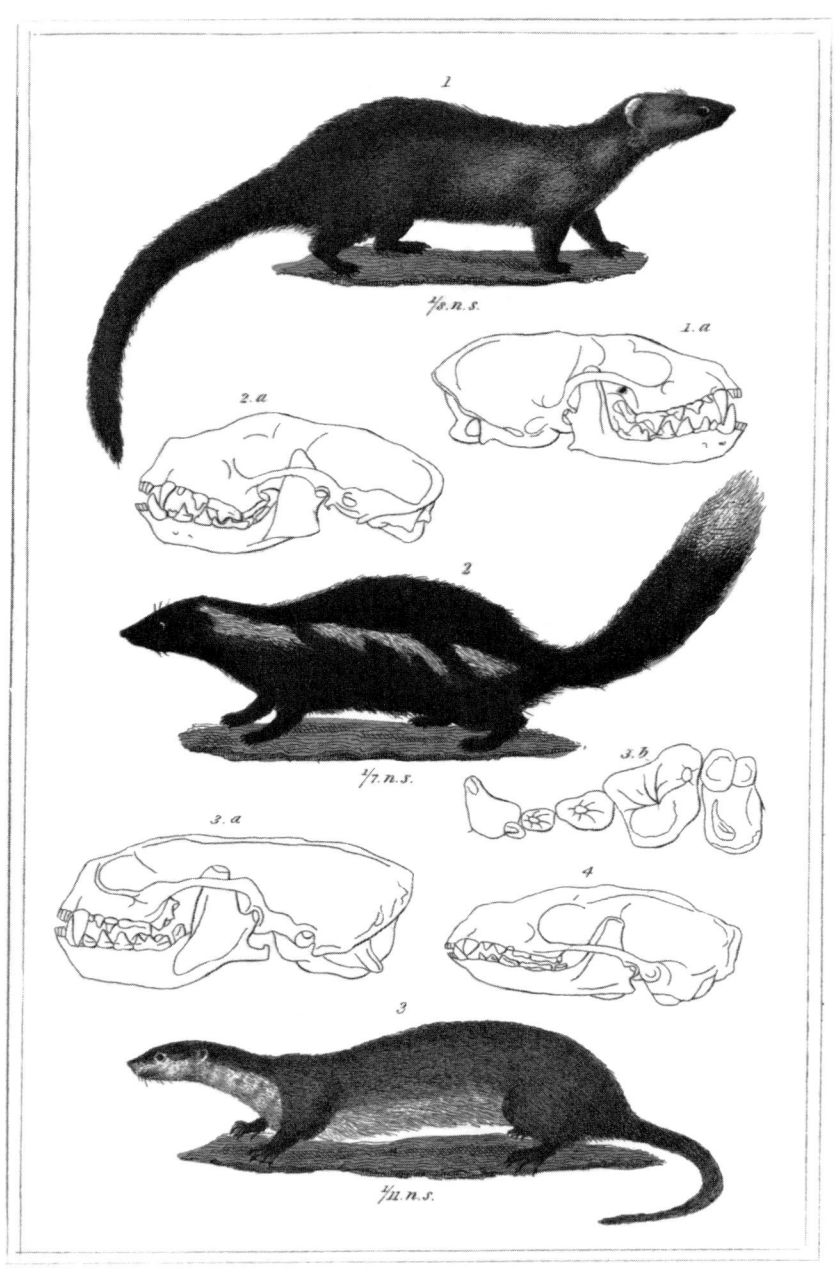

Marten, skunk, and otter, all Mustelidae and relatives of the badger

The most primitive of all meat-eaters are members of the Mustelidae family. There are about seventy species in the family, all composed of small- to medium-sized carnivores having long bodies, short legs, and, usually a short tail. The mustelines have well-developed anal scent glands which certain species—notably the skunks—use for defense. The stench of the fluid ejected from these scent glands and the bloodthirstiness of some species have given the Mustelidae a bad reputation. Nevertheless, the fur of several mustelines is highly prized.

Members of the Mustelidae are: badger, ferret, grison, marten, mink, otter, polecat, ratel, skunk, tayra, weasel, wolverine, and zoril. The weasel's Latin name, *mustela*, is the source of the word Mustelidae.

Fossil finds reveal that present-day mustelids, which have links to the dog tribe, are similar in many respects to the first-known terrestrial carnivores. Some authorities maintain that the Mustelidae might have been the progenitors of all other modern carnivores. Other experts disagree. They claim that the mustelines broke away from other carnivores at a very early date. But one thing is sure—the Mustelidae have a long family history.

One or more species of mustelids can be found in every part of the world with the exception of frigid Antarctica and isolated islands such as Australia, New Zealand, and Madagascar. Thousands of years ago, the mustelids evolved a wide variety of adaptions that enabled them and their living descendants to thrive under most climatic conditions and to prosper in many different environments. As a result, mustelids make their homes in trees, on and in the ground, and in the water. Incidentally, there is as great a range in the size of the various species of mustelines as there is in their habitats. For example, the sora, a South American otter, has a body length of six feet, while the least weasel is so small it can scurry through a hole only one inch in diameter.

It has not been an easy task for paleontologists (experts in the identification and dating of fossils) to trace the development of the mustelids. Because the common ancestors of all present-day mustelines were relatively small animals with light and fragile bones, fossil remains of early members of the Mustelidae family are usually fragmentary. In fact, few complete skeletons have been found. Nevertheless, paleontologists have established that the Mustelidae originated in the Old World. The "record of the rocks" has also revealed that most of the various types of mustelids ranging the Earth today existed by Miocene times. Actually, there is but slight difference between the Eurasian badger (*Meles meles*) and certain badgers that lived some twenty-five million years ago.

There are eight species of badger. One of them, *Taxidea taxus*, popularly known as the American badger, is found in Mexico and North America. The other seven species are native to Europe and Asia. The ranges of certain of these species are separated by hundreds of miles; the ranges of others overlap. Generally speaking, the habits of different species sharing the same habitat vary considerably.

The American badger. Some naturalists think the facial markings of both the American and Eurasian species are "warning coloration"; others disagree.

The majority of Old World badgers have a characteristic pattern of black and white stripes on the face. A number of authorities suggest that these markings are the source of the animal's common name, likening the stripes to a badge. Other experts maintain that the word badger was derived from the French *becheur* (to dig). There is also a debate as to whether or not the facial stripes are a warning coloration designed to advise other animals that the creature so marked is capable of putting up a savage defense. Those who do not think badgers have warning coloration point out that badgers are active at night, when most animals are unable to detect the stripes.

Meanwhile, naturalists woefully admit that we know very little about the life history of several species of badger. On the other hand, *taxus*' ways and wiles have been studied ever since the United States Government sent explorations across the Great Plains to the Rockies in the 1880's. The Eurasian badger has been observed by professional and amateur zoologists for nearly three hundred years.

Most Europeans are familiar with the badger. Rural Englishmen have a unique relationship with it. Indeed, throughout the British Isles, countrymen have a special affection for *meles*. Interestingly enough, most Englishmen call the badger "brock"—the accepted name for *meles* until the middle of the eighteenth century. Brock is derived from *broc*, the Anglo-Saxon word for badger.

" 'Tis as true as the fairy tales told
in the books."   —Goodrich

2 Lore of the Badger

*Should a badger cross the path
Which thou had taken, then
Good luck be thine, so it be said
Beyond the luck of man.*

*But if it cross in front of thee,
Beyond where thou shalt tread,
And if by chance doth turn the mould,
Thou art numbered with the dead.*

No one knows for sure how long English farmers have been convinced that there is much truth in this verse. But it has been established that seeing a badger stop to "turn the mould" (dig) has been considered unlucky in Japan since the early fifteenth century. Superstitious individuals in Japan claim that the badger is wicked and its greatest delight is bringing misfortune to humans while disguised as a young boy or as an elderly Buddist monk.

While the Japanese maintain that it is impossible to identify badgers that have assumed human form, the natives of Mada-

While some peoples consider badgers to be "shy," others regard them as wicked and malevolent.

gascar insist it is easy. All one has to do is look at a suspect through the left sleeve of a coat. If he is a badger, his disguise will vanish.

But the Ainu, a primitive people who live on four islands in the northeastern part of the Japanese archipelago, wouldn't even attempt to unmask a masquerading badger. They are too afraid to chance making it angry. They are sure that, if they did, the animal would seek malevolent and diabolical revenge.

Other peoples beside the Ainu consider the badger a wicked creature. In fact, this conviction is widespread throughout Europe. In most European countries, the credulous believe that the badger engages in its malicious activities under the supervision of Satan. This supposed close alliance between the badger and the Devil is confirmed by an old tale. It relates how two hunters captured a live badger, placed it in a stout sack, and set out for home. They soon discovered that the sack was weightless. Curious, the two began to untie the sack but were overcome by the stench of brimstone. Upon recovering, they found the sack empty and realized that the Prince of Darkness had released the badger.

Despite this legendary alliance between the Devil and the badger, it is widely held that the animal's fur has the ability to break spells cast by sorcerers, overcome the wickedness of witches, and render wizards powerless. Thus, early books containing formulas for thwarting the forces of evil always recommended badger hair to counteract black magic. One of these volumes assures its readers that: "A tuft of hair gotten from the head of a full-grown Brock is powerful to ward off all manner of witchcraft; these must be worn in a little bag made of cat's skin—a black cat—and tied around the neck when the moon be not more than seven days old and under that aspect when the planet Jupiter be mid-heaven at midnight."

Today, some Spaniards wear a badger's paw mounted in silver to protect them from the evil eye. But the residents of southern Italy claim that bundles of badger hair are better amulets for this purpose. Other Italians disagree. They maintain that the best defense against the machinations of witches and other evildoers is a charm made of badger skin. Meanwhile, many who live on the slopes of the Apennines—the mountain range that runs through central Italy—consider a *piccino* (a tuft of badger hair enclosed in a small gold case) an ideal Baptismal gift. Supposedly this present will protect an infant from witchcraft, black magic, and the evil eye.

Certain inhabitants of the valleys cutting through the Apennines not only rely on badger hair to shield their children from harm but also to guard their draft animals. These individuals attach a talisman fashioned from badger hair and red flannel to the yokes and collars of their oxen and horses. A similar custom was observed in the Low Countries. However, the Belgian and Dutch farmers who fastened badger skins to harness insisted the skins were merely ornaments. No one was fooled by their protestations. All knew that the object of the skins was to keep disease, injury, and spells from the draft animals.

Unlike most other European peoples, the Germans neither credited the badger with supernatural powers or linked it with the Devil. German folklore depicts the badger as a gentle, peace-loving, human-like creature whose main interests are his comfort, family, and home. But if the badger of German legend is alarmed or its usual routine is disturbed, it becomes furious.

It is usually the badger's cousin, the sly and dishonest fox, that causes the good-natured badger to lose its temper. Yet the badger is fond of his erring relative and tries to reform the fox. Although the efforts are in vain, the badger stoutly defends the fox against its many enemies. Instead of being grateful, the fox plays one malicious trick after another on the badger.

German folklore may hold the badger up to gentle ridicule but it also credits it with having the ability to forecast the weather. Tradition holds that if the badger comes out of its den on February 2 and sees its shadow, the animal will go back to its den and there will be six more weeks of winter weather. But if the day is cloudy and the badger cannot see its shadow, it will stay above ground, indicating that spring will come soon.

The celebration of Groundhog Day in the United States was inaugurated by German immigrants who transferred the legend of *meles'* ability to prognosticate the weather from the Old World badger to the groundhog (woodchuck) of North America.

Romanians maintain that vampires would vanish from the Earth were it not for the badger. They say the blood-sucking ghosts cannot climb out of their graves and wander to and fro in search of a victim unless they first smear a special ointment on their bodies. The main ingredient in this salve is badger fat.

Actually, badger fat has long been esteemed as a healing balm in Europe. The fat is said not only to have outstanding beneficial properties but also to possess great penetrating power when applied to bruises, cuts, and lame joints. British shepherds hold that if badger fat is rubbed into the back of the hand, it will pass through to the palm. Similarly, old-time American trappers who employed badger fat to limber rheumatism-crippled fingers swore it was so strong that, when applied to a head wound, it made the hair stand on end!

In some parts of Europe, badger fat is still being used to alleviate minor injuries to both humans and livestock. But the use of this fat is not nearly as widespread as it was some eighty years ago when *Sporting Magazine,* a British publication, informed its readers: "The flesh, blood and grease of the badger are very useful for oils, ointments, salves, and powders, for shortness of breath, the cough of the lungs, for the stone, sprained sinews, colic, etc. The skin being well-dressed is very warm and comfortable for ancient people troubled with paralytic disorders."

Today, the natives of Java and Borneo mix small pieces of badger skin with water and drink the potion to cure fevers. This "prescription" is far less complicated to compound than one recommended to asthma sufferers in Mexico a century ago. It called for stewing baked badger meat with flour, sugar, and pinches of various herbs until the liquid was thick. But, irrespective of how long the stew was boiled, this nostrum would not help the patient unless taken during the first quarter of the moon. If imbibed then, the patient would be able to breathe without difficulty before the moon waned.

Besides being employed as a remedy for various ills, the badger has played a minor role in preventive medicine. A hundred years ago when a bite from a rabid dog meant a horrible death, the residents of Somersetshire in southeastern

England had no fear of rabies. They believed the badger snouts they carried insured that they would suffer no harm if they were bitten by a "mad dog."

Although the Creeks, a confederacy of tribes that formerly lived in Alabama, Georgia, and parts of northern Florida, rarely saw a badger, they believed it was responsible for humans having cramps. The Creeks were convinced that the badger suffered constantly from a stomachache and, being a spiteful creature, passed some of its chronic pain on to mankind.

Actually, the redman regarded the badger with mixed feelings. Some tribes hated it; others esteemed it. Tales told by the Sioux and other Great Plains Indians are crammed with details of the animal's evil deeds. Conversely, the Shasta of California recount how the badger helped people and animals escape capture by a vicious monster. In these stories, the badger digs two holes and hides those he is protecting in one of them. When the monster asks where his intended victims are, the badger points to the other hole. As soon as the monster enters the empty hole, the badger fills it with dirt and smothers the evil one.

No Indians hold the badger in more esteem than those native to the Southwest. Most of these tribes believe that their ancestors originally lived in the Netherworld. The Hopi, Mohave, Zuni, and others are convinced that the "Ancient Ones" never could have pierced the Earth's crust and reached this world without the help of the badger.

Generally speaking, the "emergence" myths told by tribal storytellers show little variation so far as the badger's role is concerned. However, a few accounts are most colorful. One of these is the Jicarillo Apache tale. It vividly describes how the badger dug a pathway to the Upper World for the people through layers of mud—which is the reason a badger's legs appear to be stained with soil.

Another version of the coming of humans to the present-day Earth not only gives credit to the badger but also explains why it lives in a burrow. This legend from the Acoma pueblo in New Mexico states that the first two females lived underground. Tradition maintains that the gods gave them two baskets containing seeds and models of animals. The gods instructed the women to plant the seeds and breathe life into them once they found a way to reach light. The chances are the humans would have remained in darkness forever if they hadn't transformed the model of the badger into a living animal. It excavated a path to the light. The grateful females rewarded the badger by giving it a burrow with a temperature controlled by magic. Thus it was never too hot or too cold.

Because of the traditional tie between their ancestors and the badger, southwestern tribes considered the animal sacred and included it in various religious rites. Moreover, in Zuni communities, the Badger Clan plays an important part in certain religious festivals. Among the duties of this clan is caring for the holy fire during the ten-day celebration of the coming of spring.

Convinced that the badger had the power to aid humans, a number of tribes appealed to it for help. Thus, before going out to hunt, the Pima would sing a song asking the badger to "bring the feet of the black-tailed deer" toward their camp.

The Menomini Indians of Wisconsin say that the badger helped the redman develop lacrosse. But badger baiting, not lacrosse, is the sport that Americans and Europeans associate with the badger. However, no true sportsman would consider badger baiting a sport.

In its simplest form, badger baiting consists of pitting a badger against a succession of dogs. The badger, either fastened to the ground by its tail or chained to a barrel, is forced to fight until it finally dies from wounds or exhaustion.

Often, to make these horrible contests "fair," the powerful lower jaw of the badger would be broken. This prevented the unfortunate animal from securing a tight grip on its opponents. This practice was more common in England than in the western United States, where dog-and-badger fights were staged in frontier towns. In his *Adventures in Apache Country*, published in 1869, John Ross Brown describes a badger baiting he saw in Texas. Brown reports that the badger fought relays of dogs for two hours. It won every round and, as a reward, was beaten to death with a club. Brown ". . . turned away with a strong emotion of pity—there was something about the whole business very much like murder."

Badger baiting is a thing of the past. But it is recalled by the verb "to badger," meaning to annoy or to persecute. This is a complete inversion of the original meaning, because it was the badger that was "badgered"—never was it the aggressor.

Popular speech has not neglected this animal. However, while "black as a badger's back" is easily understood, "to be as uneven as a badger" requires explanation. This phrase stems from the once widely spread belief that the badger's legs, being shorter on one side than the other, are especially adapted for walking on hillsides. Another obscure expression featuring the badger is "badger game." This is a cant term used by criminals to describe the blackmailing of a man by a woman.

No one wants to be the victim of a badger game. Nor does one wish to be told that he "stinks like a badger." This idiom refers to the vile-smelling fluid a badger can eject from its anal glands.

Evil badgers, traditionally "killers of cattle and men," were once thought to live throughout Ireland. However, the badger has not given its name to many places in Erin. It is different in England. Among the dozens of geographical features and towns named for the animal, there are: Brockenhurst, Brock-

Long-bodied, short-haired dogs hunting badgers are shown in pictures painted in the fifteenth century. These dogs were the ancestors of the Dachshund, whose name, translated from the German, is "badger hound." The specimen shown here is the smooth-haired variety. There are also wire-haired and long-haired Dachshunds. Over the years, miniatures in each coat have been developed to hunt in sett tunnels too narrow for larger Dachshunds to enter.

hampton, Brocklebank, Brocklands, and Brockholes. Moreover, the English name of Brockman denotes that the original members of this family were professional badger hunters.

Dach, the German name for badger, appears in such place names as Dachsbach, Daschberg, and Dachesfelden. Besides having given its name to these locales, the badger has been memorialized in another way in Germany. In the early fifteenth century, hunters seeking a dog courageous enough to enter a badger's burrow, squeeze through its passageways, and battle the cornered inhabitant began the selective breeding of long-bodied, short-legged, loose-skinned dogs. The result of their efforts was the dachshund or badger dog.

Like its European kin, the American badger has given its name to numerous towns, and such place names as Badger Creek, Wisconsin, mark the presence of badgers in that vicinity. The area where Illinois, Iowa, and Wisconsin meet is known as The Land of the Badger. Meanwhile, Wisconsin's nickname is the Badger State. According to an article in the Madison *State-Journal* for December 10, 1879, the ". . . term 'badger' was first applied to the early lead miners, who on first coming to a new location dug into the side of a hill and lived underground like badgers. In time all the inhabitants of the lead-mine region became known as 'badgers' [and] not by an unnatural adaption it has been applied to the State itself."

Respect for the badger's courage and fighting spirit has led a number of institutions of higher learning to call their athletic teams the "Badgers." Among these schools are: Amarillo Junior College in Amarillo, Texas; Pacific University, located in Forest Grove, Oregon; and Spring Hill College, Alabama.

The badger has been burrowing through literature for centuries. Long before Shakespeare, Mark Twain, Horace Walpole, and other authors wrote about the badger, it was mentioned in the Bible. One very early edition of the New Testament renders Hebrews 11,37 as "thei wenten about in brock skynnes." A covering of badger skins is cited in the directions for building the tabernacle given the ancient Jews (Exodus 26,14). We know that some Israelites were "shod with badger skin" because of a description of their dress in Ezekiel 16,10.

Not only is the badger the leading character in fables told throughout Europe but also it is the central figure in many Japanese tales. In some of these stories, the badger is always outwitted by the fox. In others, the evil ways of the badger are stressed, and in one famous yarn, "Kachi-Kachi," a badger commits murder and is brought to justice by a hare.

Perhaps the best-known poems about the badger are Jean Clare's "The Badger" and Robert Grave's "The Six Badgers." But far more popular than these poems are Beatrix Potter's charming accounts of the activities of Tommy Brock. Russell Hoban's more recent books tell of the adventures of Frances, one of the most engaging animals to live between the covers of a book.

A 1972 Newbery Honor Book, *Incident at Hawk's Hill* by Allan W. Eckert, is about a boy adopted and protected by a badger. Set near Winnipeg, Canada, the novel is based on a true incident of the 1870's. At once heartwarming and heartbreaking, it attests to the author's knowledge and understanding of wild creatures—be they boys or badgers.

Master painters have found little inspiration in the squat badger, whose body is wider than high. However, both the Old World badger and the American badger have been depicted in sporting prints, engravings, and natural history books. Some of these representations of the badger are the work of such famous artists as Thomas Berwick of England and John James Audubon of the United States.

The American badger by Audubon

Because few masters of the brush or craftsmen have been inspired by the badger, the finding of this sculpture of a Eurasian badger and cub in an antique shop in Switzerland was a most unusual event.

While art connoisseurs avidly seek delineations of the badger that bear the names of internationally famous artists, animal lovers and collectors of modest means purchase badgers fashioned by craftsmen from clay, glass, metal, and wood. The chances are that, in time, some of these delightful works of art will be as highly regarded as are the creations of Berwick and Audubon.

Professional gamblers who made their living playing cards with cowboys, miners, and mule-skinners in the Old West would be amazed to learn that books have been written about the badger. To all residents of frontier towns, the badger was a varmint—a horse could break a leg in a badger burrow. However, despite their dislike for the animal, most old-time card-sharps sewed a badger's tooth into the lower right-hand pocket of their vests. They were sure that the tooth would bring them luck.

"Nature has given us two ears and one mouth." —*Disraeli*

3 *Physical Characteristics*

WHEN Carl Linnaeus (1707–1778), the famous Swedish botanist, classified plants and animals, he placed badgers in the same group as the bears. This was a mistake. Badgers and bears are not related. However, both the Eurasian and American badgers resemble small bears as they waddle through the night, seeking food.

Their likeness to miniature bears is not the only similarity between *meles* and *taxus*. They have many physical characteristics in common. There are also several differences between the two. This chapter describes those features the Eurasian and American badgers share and also calls attention to their dissimilarities. Before these comparisons are made, some pertinent facts about *meles* and *taxus* are presented. A later chapter deals with the various exotic species of Old World badgers.

As indicated, *meles*' family history is a long one. Eurasian badgers were excavating setts (the technical term for badger dens) long before man learned to make fire. Today, the Eurasian badger is distributed throughout the greater part of Europe and Asia, including Japan. The northern limit of *meles'* range is just below the Arctic Circle, while none live south of the Himalayas. *Meles* is common in some parts of its range. In other places, it is rare.

Some authorities maintain that there is only one species of Eurasian badger. Others claim that there are a number of subspecies—with no agreement as to how many. Still others hold that most of the groups of badgers their fellow scientists classify as subspecies are races (individuals with a number of common attributes). However, all zoologists believe that isolated groups of Eurasian badgers—such as *Meles meles rhodius* found only on the Island of Rhodes in the Aegean Sea—are deserving of subspecies rank.

Both male and female Eurasian badgers have stocky, wedge-shaped bodies that measure up to three feet from tip of nose to end of tail. The tail itself is a stump, being approximately four inches long. Depending upon the season—badgers are fatter in the fall than in the spring—*meles* weighs from twenty-five to thirty-five pounds. Although there are numerous records of

The Eurasian badger is chunky with powerful muscles and a loose skin.

boars (males) weighing forty pounds or more, the average Eurasian badger's weight is twenty-seven pounds. Sows (females) are a few pounds lighter and slightly smaller than boars.

Because the Eurasian badger is chunky, it looks smaller than it really is. With the exception of deer and the free-roaming ponies of the New Forest, *meles* is the largest wild animal inhabitating the British Isles. For its size, the Eurasian badger has tremendous strength. Like all mustelids, badgers have inherited the powerful muscles of the carnivore's ancestors. Their stocky bodies are covered with a skin "a size and a half too big" and so loose that the badger "can almost turn around inside it." When a dog grabs a badger at the neck or back, all it grips is a mouthful of skin.

Originally, *Taxidea taxus* was distributed from southwestern Canada to Mexico and from Illinois to the Pacific. Once a dweller on the open plains, *taxus* has recently taken up residence in farmlands or on the fringe of wooded areas. As a result, sightings of badgers have been reported as far east as the State of New York.

Fossil remains of a badger dating from Pleistocene times found in Maryland's Cumberland Cave are most likely those of *taxus*' immediate ancestor. Zoologists recognize this extinct species and four living subspecies of *taxus*.

According to naturalist Lugwig Heck, the American badger "looks very turtlelike, much as if someone took a half-grown Eurasian badger and beat it until it was broad and flat." Rarely three feet long, with a four- to six-inch tail, the average American badger weighs between eight and twenty-six pounds.

Bowlegged and pigeon toed, with a body wider than it is high, *taxus* is not the most attractive of animals. But its short, flat, thickset body and its short legs powered by very strong muscles are ideal for a fossorial (burrowing) animal.

Even stockier and usually smaller than the Eurasian badger, the American species appears to be flatter. Its black and white facial stripes are narrower and less distinct. Both species are heavily built animals whose muscular legs seem even shorter than they are.

Fur

Both American and Eurasian badgers have such thick fur that they can spring a trap by rolling over on it and leave nothing behind but a tuft of hair. However, *taxus* has much softer pelage than its Old World kin. *Meles*' ill-fitting skin is covered with long, thin, stiff, and coarse guard hairs and scanty underfur.

All of *taxus*' guard hairs have a narrow black band just below their white tip. The color of the American badger's fur coat ranges from a grizzled silver gray to reddish-brown on the upperparts, while the underparts—except for the brown or black legs—are buffy. A white stripe extends from the pointed, slightly upturned snout at least to the shoulders. Cheeks, chest, and throat are white.

Seen in dim light when the wind is blowing strong enough to ruffle its fur, a Eurasian badger appears to change color. One minute it looks white, a moment later it takes on a blackish

An American badger at the entrance to its burrow. Badger fur tends to lighten with age.

sheen. This shifting of colors enables *meles* to merge into the shadows and, at times, become invisible.

Actually, *meles'* fur does not change color. It only seems to do so because of the way the hairs are pigmented. The upper third of each hair is white, the middle third black, and the bottom third white down to the roots. Thus, if a stiff wind disarranges the fur, an onlooker gets the impression that the pelt changes hue. Meanwhile, seen from a distance in moonlight, *meles* looks as if it were dressed in grizzled gray.

Badger fur may be permanently soil stained, the result of years of burrowing. Both *meles* and *taxus* clean their fur frequently, using their foreclaws as combs. Albino (pure white) and melanistic (all black) badgers are relatively common. However, not all white badgers are albinos. Badger fur has a tendency to lighten with age. Thus the coats of older individuals are often whitish.

The outstanding feature of *meles'* coloration is the facial mask. It is composed of two rather wide black stripes on either side of the face extending from the back of the small, white-tipped ears, over the eyes, almost to the tip of the tapered muzzle. Bordering bars of white run from the snout to the top of the head.

Taxus lacks a facial mask. But it has patches of black surrounded by white fur before and behind the ears. These markings are known as badges, and some authorities claim they are the source of the word badger. Other experts theorize that the well-defined black and white stripes on *meles*' face serve as camouflage, "simulating moonlight coming through the trees." This theory is flawed. The Eurasian badger avoids moonlight and also, when *meles* does show itself, it is very conspicuous.

Legs and Feet

The strength of the badger's short legs is out of all porportion to its size. The forefeet are used as spades and pickaxes, being armed with formidable compressed and curved claws. Because badgers constantly dig, it is not unusual for the claws of old animals to be worn down.

Both *meles* and *taxus* can excavate a burrow faster than a man can dig a ditch of the same length. Badgers prefer to dig in loose sandy soil, but their extremely strong claws and muscu-

Badgers are prodigious diggers and they can send the dirt flying.

lar legs enable them to bore through compacted soil and to move good-sized rocks out of their way. When burrowing, badgers loosen soil with the curved fore claws and send it flying with the shovel-like hind claws.

Besides using their claws to dig and to hold and tear prey, badgers employ them to relieve itches and flea bites. Badgers give themselves a "pedicure" regularly to remove dirt lodged between their claws. Interestingly enough, although badgers dislike getting their paws and claws wet, they enjoy bathing. The American badger not only swims and dives but also sprawls in shallow water on a hot day to cool off.

Like man, the badger is plantigrade. This word, derived from the Latin *planta* (sole of the foot) and *gradior* (to walk), is used to identify animals that walk on the whole sole of the foot. A badger's gait depends upon circumstances. When alarmed, it will quicken its shambling pace; if it feels secure, it moves slowly. When engaged in a specific activity, badgers usually move at a fast amble, head down, hindquarters swaying, short legs pumping furiously. But, unless scurrying to the safety of its sett, the Eurasian badger normally makes frequent stops at irregular intervals and listens for sounds indicating possible danger.

Badger footprints show the five toes on each foot. The foreprints are wider than the prints of the hind feet. Width of the

Female badger's forefoot

Male badger's forefoot

Female badger's hind foot

Male badger's hind foot

Not only is the print of a badger's forefoot larger than that of the hind foot but also, when a badger runs, the hind feet slightly overlap and imprint the forefeet. However, this may not occur when the animal moves slowly.

prints indicates the size of the animal that made them. The wider the prints, the bigger the badger. Prints of the hind feet slightly overlap those of the forefeet.

Walking backward for long distances is no problem for *meles* and *taxus*. *Meles* makes the most use of this ability. When hunters send a dog trained to enter setts and harass—but not injure—the occupants and keep them from escaping, *meles* will face its tormentor. Slowly backing up, *meles* leads the dog deeper and deeper into the labyrinth of passageways that wind through a sett. Above ground, the hunters, guided by the dog's barking as well as a bell attached to its collar, start to dig.

Hunters often shovel for hours while *meles* continues its strategic retreat. An English magazine published over a century ago described a badger hunt during which four men dug in relays for two days along a burrow eighteen feet below the surface! There are times when hours of digging and "earth stopping" (plugging sett tunnels with soil) fail to unearth a badger. Moreover, dogs have been known to enter a sett on one side of a hill and emerge on the other side a day or two later.

When the American badger is threatened near its burrow, it faces its foe and slowly backs into its den. Once inside, it blocks the entrance with soil. If cornered away from its burrow, *taxus* makes no attempt to flee. A poor runner, the badger begins to dig, showering its attacker with dirt as it quickly vanishes.

It is practically impossible to lift a living badger out of one of its tunnels. The animal braces its feet and refuses to budge. When relocating badgers because their population is too high for the available food supply, English badger hunters overcome *meles*' stubborn resistance with a specially designed set of tongs. Not only do the tongs supply enough leverage to pry a badger loose but also they greatly reduce the danger of being bitten. Incidentally, grabbing a badger from behind does not insure safety—the animal may twist around suddenly and sink its teeth in the hunter's hand.

Teeth

For their size, badgers probably have the most formidable set of teeth of any animal. Moreover, when badgers bite, their

Badger teeth are formidable, as this snarling specimen attests.

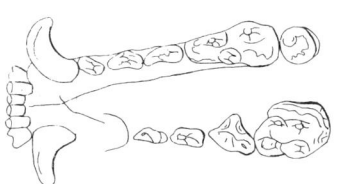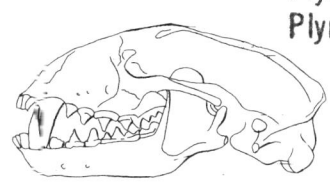

The jaw and skull of a badger, drawn in 1707

jaws lock. As a result, the Eurasian badger's thirty-eight teeth and the American badger's thirty-four are capable of inflicting a ghastly bite. It was formerly believed in central Europe that once *meles*' teeth closed on human flesh, the animal would not let go until it heard the crunch of breaking bones. Therefore, Danish badger hunters wore two pairs of shoes separated by a layer of charcoal. Presumably, if a badger bit a hunter's foot, it would release its grip when it heard the charcoal crumble.

Badger dentation differs from that of other carnivores. Although carnassials (teeth adapted for stabbing) are present, they are not as developed as those of most flesh-eaters. Then, too, the badger's molars are rather flat so the animals can grind vegetation easily. The canines (tearing teeth) are more prominent than the incisors (cutting teeth).

Not only do badgers' teeth seem designed for attacking and eating large prey but also the jaws are exceptionally powerful, being moved by very strong muscles. As a matter of fact, badgers have the same type of jaw as do those mustelids that prey on animals larger than themselves. Yet badgers limit their hunting to small birds and rodents, and most of their menu consists of soft foods.

If the flesh is removed from the lower jaw of most animals, the jaw falls away from the skull. But the badger's lower jaw is so hinged that the skull must be fractured before it can be dislocated. The skull itself is extremely hard. If a badger is hit on the head with a gun butt, the chances are it will stop, shake its head several times and shuffle away. But badgers cannot survive a relatively light tap on the nose.

Senses

Because badgers are more at home in darkness than in light, it is logical to assume that they have excellent night vision. Actually, they have poor vision, day or night. Zoologists doubt that badgers can perceive colors, believing they see all objects as either black or gray. Nevertheless, badgers are quick to note the slightest movement within range of their vision. This is not very far on a cloudy night and only about fifteen feet on a clear one.

Wary creatures, badgers memorize the physical features of their territories. They know the location of every bush and stand of ferns that sway in the wind—and pay no attention to them. But unfamiliar movements are quickly detected and prompt an immediate reaction. Conversely, badgers do not associate stationary objects with danger. If a human stands still and there is no wind to blow scent, an approaching badger will usually pass close by in the darkness.

Although badgers' external ears are small (an adaption for life underground), both *meles* and *taxus* have acute hearing, being able to pinpoint the rustle of leaves or the breaking of a twig a considerable distance away. Usually alert, badgers stuffed with food are lax at times and do not listen for suspicious sounds. Confident that they have nothing to fear, they plod along making a great deal of noise. Under these conditions, it is possible to get quite close to a badger, providing it does not pick up one's scent.

Before setting out for a night's foraging, badgers sniff in all directions to "test the wind." Their keenest sense is that of smell, which is so sharp that badgers can detect a man's scent long after he has walked over an area. If a Eurasian badger smells

Badgers rely on their noses to warn them of danger. This early representation elongates the nose considerably.

a man, it scurries toward the safety of its den, while an American badger will burrow out of sight. If a cub is handled by a human, its mother will kill it. Incidentally, although badgers rely on their noses to warn them of danger and to locate food throughout the year, zoologists have learned that a badger's sense of smell is keenest when it is raising a family.

Scent Glands

Anal musk or scent glands are characteristic of the Mustelidae. Most members of the family use the fluid ejected from their glands to defend themselves, mark territories, foul food so no other animal can eat it, or attract mates. While *taxus* may expel—but not spray—its musk mainly when fighting, *meles'* glands are stimulated by excitement. Thus the Eurasian badger's glands are as likely to be activated when it is playing as when it is frightened.

Whether badgers deliberately or accidentally discharge the nauseating secretion produced by their scent glands, the fluid gives the animals their distinctive odor.

Voice

Taxus is a much quieter creature than the Eurasian badger. While the American badger makes a "purring" sound during a successful courtship, it rarely uses its voice except when forced to fight. Then it will face an adversary and attack with ferocity, growling, hissing, snarling, and squealing.

Meles is very noisy. It sniffs constantly to pick up scents, and emits the same sounds as the American badger when battling for its life. Like its New World kin, the Eurasian badger relies on a long, drawn-out purring to attract mates. Sows employ a variation of the purr to keep cubs inside the entrance of a sett when the babies are too young to go above ground.

Besides a repertoire of squeals, whines, howls, and other cries, *meles* lets out an eerie scream. This weird cry, half whisper, half shrill howl, is so unearthly that British zoologist Brian Vesey-Fitzgerald admits that when he heard it the hairs on the nape of his neck "stiffened involuntarily."

When playing together, Eurasian badgers yelp excitedly. The more intense their roughhousing, the louder the yelps. But the yelps are never as high pitched as the squeals with which the cubs welcome their parents home from a night of hunting for food. Undoubtedly the most unusual cry made by the Eurasian badger is the piercing scream of males during the mating season. It is ". . . not unlike that heard from an injured man or woman. This call resembles the death scream of a mortally wounded badger, a sound which is so terrifying that many a hunter has ceased getting badgers after hitting one and hearing it cry."

"Nothing is done without a reason."
—*Seneca*

4 Badger Behavior

NOT only do *meles* and *taxus* share certain physical characteristics but also they have habits in common. Nevertheless, in some respects, the life-styles of the American and Eurasian badgers differ. For example, except during the mating season, male American badgers are solitary, and sows live alone save when raising a family. On the other hand, *meles* delights in the company of others of its species throughout the year. There is also a variation in the way *meles* and *taxus* set up housekeeping.

Badger Barracks

No animal is more particular about its "address" than the Eurasian badger. It may pick a homesite on the edge of a forest, along a thick hedgerow, on the sides of a deserted sand pit, in an abandoned quarry, or in a sunny, wooded area. But it prefers to live in a hilly, wooded spot where the soil is light and well drained, near farmland, and not too far from a source of drinking water. It dislikes wet lowlands and, although found 7900 feet above sea level in the easternmost reaches of its range, it usually avoids mountainous regions.

As indicated, the badger is "pound for pound the animal

A young Eurasian badger about to enter its sett. Note the mound of dirt around the entrance. Although the sett is a new one, its occupant has evidently tunneled a considerable distance underground.

kingdom's most efficient digging machine." Eurasian badgers are diligent miners, spending much of their lives scraping the walls and floors of setts and boring new tunnels. Kenneth Grahame, in his charming *The Wind in the Willows*, has a badger explain its compulsion to dig. According to the badger: "There is no security, or peace or tranquillity, except underground. And then, if your ideas get larger and you want to expand. . . . why a dig and a scrape, and there you are!"

If not disturbed by man, successive generations of Eurasian badgers will occupy the same sett for a century. As the decades pass, the animals honeycomb the ground with connecting tunnels at various levels. These passageways, which may extend a hundred feet or more, are "air conditioned." Small holes in the roofs of the tunnels nearest the surface serve as vents. If one of the topmost tunnels collapses, the dirt is removed and the opening becomes a new entrance hole.

Long established setts may have forty to fifty entrances. The location and number of those in use depend upon the season. Unused openings are employed as emergency exits. Main entrances in ancient setts run to a large chamber, the floor of which has been worn smooth by the feet of hundreds of badgers. These chambers—probably the nurseries excavated when the setts were first dug—are called ovens. Some ovens are so large two men can sit comfortably in them.

Even if two or three pairs of Eurasian badgers inhabit a sprawling sett, they cannot occupy all the chambers it contains. Entrances, passageways, and rooms not used by the sett's owners are often subleased by foxes. *Meles* dislikes the scent of its uninvited guests and usually blocks off the area where the foxes live. Chambers in which old or injured badgers have died may also be sealed. Years later these chambers may be reopened and the dead animals' bones kicked out to lie amid the stones and dirt strewn around the sett's current entrances.

Dried leaves, ferns, grasses, straw, and other vegetation are also scattered in front of setts. Plant material is strewn along the paths leading to the sett, having been dropped by badgers collecting bedding. *Meles* gathers bedding with the front legs, then either pushes it into the sett with the nose or holds the bedding with the forearms and chin while walking in backward. Not only is bedding changed from time to time but also it is aired regularly by being nosed to the surface and spread out to dry. Eurasian badgers are fastidious housekeepers. Each spring, setts get a thorough cleaning. Badgers never foul their dwellings. Within a few yards of every sett, there is a shallow pit for use as a "latrine."

The burrows dug by the American badger in sandy knolls, along ridges, in open grasslands, and in desert sand enable it to withstand extremes of heat and cold. *Taxus* thrives in California's Death Valley two hundred feet below sea level and in alpine meadows at elevations of ten thousand feet or more. But compared to the setts of the Eurasian badger, the burrows of *taxus* are merely holes in the ground. There are two reasons for this. The first is that *taxus* does not occupy the same sett

The American badger seems to prefer digging a prairie dog out of its burrow to digging a burrow of its own.

generation after generation. Secondly, the American badger seemingly prefers to spend its time digging rodents out of their holes rather than boring tunnels. Nevertheless, some of *taxus'* excavations are deep and run quite a ways underground. Entrance tunnels dug by American badgers thirty feet long are not uncommon. A group of naturalists in Nevada trying to capture a badger excavating in sandy soil dug as fast as they could for four hours and never caught up with their quarry. All they had to show for their arduous labor was a trench six feet deep and twenty feet long.

Although *meles* usually lives in the same sett all its life, *taxus* may dig a burrow after killing a rabbit and stay underground while consuming it. Then the badger will abandon its temporary shelter. As a matter of fact, the American badger may dig a new burrow every day during the summertime. With the approach of fall, certain of these shelters are reoccupied and, in colder parts of its range, *taxus* maintains only one den during the winter.

Each spring, female American badgers choose a suitable site and excavate a nesting burrow. Field studies conducted by Richard Lampe, a biologist on the faculty of Buena Vista College in Iowa, have revealed that although these underground nurseries vary greatly in length, all have the same features: a long entrance tunnel with an eight- to twelve-inch elliptical opening, a breeding chamber, and a safety area to which the female can flee if her burrow is invaded. When this happens, the sow digs rapidly, packing the dirt she loosens into the entrance of the safety area and blocking it. Once this "door" is in place, the female burrows deeper into the ground.

Although American badgers are often active by day in remote parts of their range where they feel safe from an encounter with man, they are nocturnal creatures. Thus it is much easier to locate the burrows than it is to spot the animals that make them. Enlarged entrances to rodent holes provide evidence that a badger is hunting in the vicinity. *Taxus*' den can be identified by its elliptical entrance surrounded, except for birthing dens, by a large mound of dirt overlaid with bits of bone pieces of fur, and rattles from snakes killed and eaten by the badger.

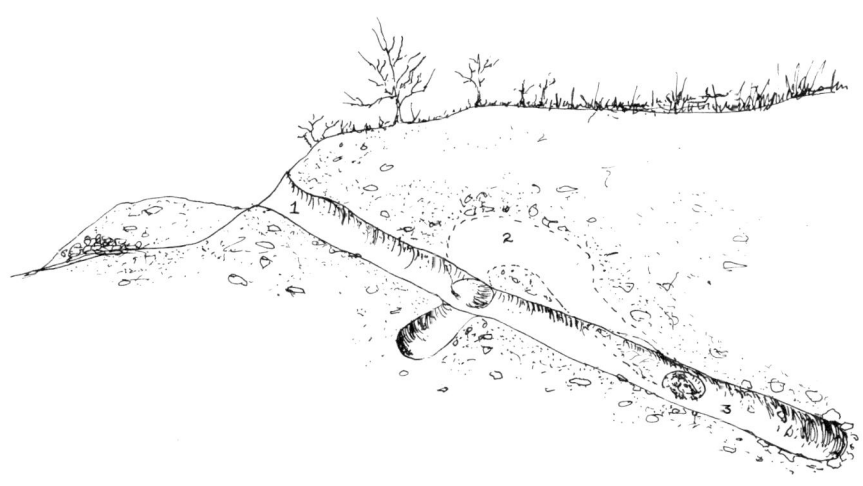

A badger burrow (after Smith): 1. Entrance; 2. Nursing area; 3. Safety area, in case of invasion.

Food Gathering

Eurasian badgers feed at night. When it is dark, *meles* emerges from its sett, squats at the entrance, and, after looking to see that all is well, goes to the lavatory pit. The next stop is the scratching post—a nearby tree where the badger stands on the hind legs, reaches up the trunk as far as possible, and brings the claws downward. After doing this for a few minutes, *meles* is ready to look for food.

However, if the badger becomes alarmed, it will plunge back into the sett. Once in the safe underground chamber, a Eurasian badger may not attempt to leave again that night. This means that it goes hungry. Food is never brought into a sett by Eurasian badgers, but they do bury food elsewhere for future use.

Once *meles* sets out to find a meal, it wastes no time. Nose to the ground, it gobbles up grubs, slugs, snails, and insects of all kinds. After a rain, Eurasian badgers forage in the damp grass, seeking the large earthworms that fishermen call night crawlers. These worms are one of *meles*' favorite foods. When looking for ground-dwelling invertebrates (animals without backbones), the Eurasian badger roots like a pig, turning the soil over to a depth of two inches. Karl von Frisch, an internationally famous authority on animal behavior, reports that after badgers rooted in the meadows near his summer home in Austria, the fields looked "as if they had been visited by wild boars."

Acorns, berries, bulbs, fungi, and grasses are important items on *meles*' menu. Before an epidemic wiped out most of Britain's rabbit population, young rabbits were a staple food of the Eurasian badger. Where the food supply is low and the badger population high, some individuals may supplement their diets with poultry, but this does not happen very often.

When *meles*' nose leads it to a nest containing baby voles (the European equivalent of the North American field mouse), its powerful forelegs and razor-sharp, two-inch claws dig out the youngsters. Legs and claws not only enable *meles* to shred rotting tree stumps and expose ants and other insects but also to rip apart bee and wasp nests. No feast appeals more to badgers than honey, bees or wasps, and their larvae and pupae. The badger's thick fur protects it from stings and it usually dines well before being bitten repeatedly on the nose or some other tender spot and forced to flee.

Some years ago, badger fur was in demand for trimming women's coats and making shaving brushes. The long hairs that were processed into brushes brought eighty-five dollars a pound. There was also a market for badger fat, which was utilized in the manufacture of salves and soaps and employed too as a lubricant for shoe leather. Therefore, in hopes of making their fortunes, a number of enterprising Americans established badger ranches. These ranches were not successful. The badgers that were kept for breeding stock ate up all the profits.

Few animals have a more voracious appetite than the American badger. It will eat any living creature it can catch and hold. If fresh meat is not available, *taxus* dines contentedly on car-

Strangely enough, this picture of badgers feasting on carrion appeared in a book for young children. Published in the late 1880's, the book describes the badger as a ferocious, blood-thirsty creature.

rion, wild plants, or cultivated crops. However, it does not consume as much vegetation as the Eurasian badger. *Meles* devours more plant material than any other mustelid, and it is not unusual for vegetation to make up three-quarters of its diet.

As noted, in areas where it has no reason to fear man, the American badger often forages early in the morning or late in the afternoon. Otherwise, *taxus* seeks ground squirrels, hoary marmots, pocket gophers, mice, rats, woodchucks, and other small game at night. Because its prey is far more agile than it is, few are captured above ground. *Taxus* usually digs prey out of burrows. However, it may enter a rodent's residence (after enlarging the entrance) when the owner is not home and wait for its return. If an American badger's hunting and foraging accumulates more food than it can eat, the surplus is cached for future use.

Taxus' menu includes the eggs and young of ground-nesting or roosting birds, baby rabbits, rodents, insects, frogs, and lizards. The American badger is very fond of rattlesnake meat and has little difficulty in securing it. Its thick fur prevents a snake's fangs from injecting deadly venom into the bloodstream. However, *taxus* cannot survive if bitten on the nose.

An American badger carrying away a rattlesnake for dinner

Coyotes often take advantage of *taxus* as it bulldozes its way into a rodent's burrow. Intent upon its digging, the badger does not notice that the coyote has sneaked close to the rodent's escape hole at the other end of the burrow. When the badger's intended victim emerges, the coyote grabs it. The wily coyote's thievery has led to the mistaken belief that badger and coyote are hunting partners.

Temperament

Although both the American and Eurasian badgers are shy, retiring creatures that would rather flee than fight, they have nasty tempers. For example, *meles* has been known to attack humans walking along the paths that lead to its water supply and hunting territory. However, such assaults are extremely rare.

While the American badger does not hesitate to bite a meddling human, it does not really become angry until cornered. Then it will face its foe, hiss, breathe heavily, wrinkle its nose, curl its lips to expose the teeth, grunt, snarl, and bark. Meanwhile, the tail is erected and the anal glands eject, wetting the long tail hairs and the animal's hind parts. A furious fighter when aroused, *taxus* makes the most of its tremendous strength. Indeed, only those who have seen an irate American badger appreciate how strong it is.

Grooming

Daily—usually before foraging—badgers clean their fur. They use not only their claws as combs but also their teeth. The teeth are further employed to remove the mites that crawl through the fur as the animals scratch

Badgers dexterously manipulate their hind feet so that they can scratch the back of the ears, the flanks, and other hard-to-reach places. When finished scratching, badgers lick their fur in the same fashion as do cats; then they pat it into place with the paws, which have been cleaned. In order to reach all parts of their bodies while grooming, badgers bend, twist, turn, and roll. Their efforts are worthwhile—badgers, unlike most long-haired mammals, never suffer from mange.

Playtime

If the weather is pleasant on a summer evening, Eurasian badgers may sprawl on the warm, bare soil scattered around their sett's entrance. Cubs, if present, do not join their parents. Instead, they play tag or king of the hill, wrestle, and turn somersaults. The more fun the youngsters have, the louder the squeals of pleasure. Then, without warning, the cubs and any adults who have joined in the games stop their rough-housing. Their hunger is stronger than the desire to play and they go looking for a meal.

Taxus' offspring also like to play. Unlike their cousins in the Old World, American badger cubs are content to bask in the sun at times, and they can be seen with their mothers close to the den's entrance on bright, warm days.

Keeping Warm

Zoologists formerly were convinced that badgers hibernate. Today, scientists agree that, in the northern parts of their range and at higher elevations, American and Eurasian badgers spend a great deal of time underground during the winter but do not hibernate. In true hibernation, an animal's body temperature drops, the muscles contract, the heart beat and respiration slow, and the need for oxygen is slight. None of these things happen to badgers, which den up during the winter months for varying periods of time.

While blizzards howl above them, badgers spend hours sleeping. Unable to hunt, they live on the layer of fat under their fur. The more they ate during the summer and fall, the thicker the layer. Most badgers accumulate enough fat to keep them in excellent condition for long periods. This has been established by various experiments. One field naturalist waited until a badger entered the only opening to its den, blocked the passageway, and set a trap at the entrance. This insured that the badger would enter the trap or starve to death. Fourteen weeks after the trap was set, the badger was captured. With the exception of being rather thin, the animal showed no ill effects from going without food for over three months. This is not a record. There are authenticated reports of Eurasian badgers in Siberia denning for almost seven months.

Incidentally, having settled the hibernation question, zoologists are investigating a strange habit of *taxus*. It is "sometimes found in a state resembling total paralysis which looks very deathlike, but if the badger is touched it runs off 'like lightning.'" Scientists are anxious to know what causes this.

Courtship

In all probability, Eurasian badgers mate for life. American badgers choose a new mate every year, and there is evidence

A mated pair of Eurasian badgers at the mouth of their sett

that some boars form alliances with more than one female. The mating season of both species of badger is from midsummer to early autumn.

Female badgers are coy. They flee from their suitors and are vigorously pursued. Eurasian sows usually run from males in a straight line, but at times they dash about in circles. First they go clockwise, when counterclockwise. This puzzles zoologists. If running in circles is an instrinsic part of Eurasian badger courtship, why isn't it done by all the females of the species?

Badger courtship is as noisy as badger play. Mate-seeking Eurasian badgers scream, and both American and Eurasian males yip and yelp with frustration as their potential mates run away. When a female allows a courting male to capture her, the yips and yaps give way to a soft purring, as noted. The boar's love song becomes a duet as the female starts to purr. Then, suddenly, the male grabs the female by an ear or the nape of the neck and the two mate.

Raising a Family

Although badgers mate from late July until early autumn, the embryos do not begin to develop until December or January. Then their growth is rapid. Biologists call this arrested development of unborn young "delayed implantation." A characteristic of many mustelids, it permits badgers and their kin to have their babies in the spring when weather is apt to be ideal for the cubs' survival and there is plenty of food.

Both *meles* and *taxus* bear their young between February and April. Shortly before the cubs arrive, Eurasian boars move to a different part of the sett. But they do not desert their mates. They regularly stand in the tunnel leading to the chamber the sow has lined with soft vegetation and gently purr. On the other hand, with very few exceptions, American boars abandon their mates by the end of August.

There are two to five cubs in a litter. The youngsters are

born blind and furred. However, not only are the white markings indistinct but also their underparts, including the inner side of the leg, are devoid of hair. Well formed, measuring between four and five inches, of which one-fourth is tail, a healthy cub weighs approximately three ounces. By the time the eyes open ten days after birth, the birth weight has doubled.

Until the youngsters are between six and eight weeks old, they remain underground, nursing, sleeping, and learning to obey their mother. Disobedient cubs are soundly cuffed. Sows train their offspring to use an excrement chamber and constantly groom them. This keeps the nursing chamber and its occupants clean.

When the sow goes hunting, the more daring of the cubs follow her to the entrance hole. Eventually, the female decides that her litter is old enough to "see the world." Sticking her head out of the burrow, she looks around, sniffs, and, if satisfied all is well, goes back for the cubs. Usually, one or two have to be coaxed outside. Moreover, even the boldest babies dash back to the safety of the nursery at the slightest noise.

Before long, the youngsters overcome their fear and are not content to play near the entrance hole. They want to go exploring. Their mother watches them constantly—female badg-

Immature American badger

Badger sows are excellent parents and, as shown in this illustration from a children's book published a century ago, will defend their cubs bravely. Note how the artist has depicted the cubs' reactions to the dachshund attack. One appears to be frightened, another curious, while the third looks as if he is going to join in the battle.

ers are outstanding parents—and trains them so well that she can leave them alone for longer periods while she hunts. Male Eurasian badgers contribute to their family's menu, but it is most unusual for an American boar to bring food to cubs.

Once the cubs are weaned, they are taken hunting and taught all the things a badger must know to survive. Sows not only instruct their young, they also play with them and shower them with affection. But when autumn comes, female American badgers drive their cubs away from the burrows in which they were born. Eurasian badger cubs are also evicted from setts in the fall. However, sometimes Eurasian badgers tolerantly allow the young of the year to remain in the sett for a while.

Eventually, badger cubs find homes of their own. If all goes well, these self-sufficient juveniles will become adults, find mates, and raise families of their own.

> "Variety's the very spice of life . . ."
> —*Cowper*

5 Exotic Badgers

NOT all the Mustelidae are as familiar to both laymen and zoologists as the American and Eurasian badgers. As a matter of fact, there has been but little investigation into the life history and habits of certain of the six species of badger that live in China, the East Indies, and other parts of the Far East. However, enough has been learned about *meles'* and *taxus'* Oriental relatives to show that they are worthy of further study.

Hog Badger

Few animals have more appropriate popular names than *Arctonyx collaris*. Because this species has a long, movable, naked, piglike snout with the nostrils on its square, truncated tip, *collaris* is called the hog badger. While rooting for food or when digging dens, the hog badger displays great skill as an excavator. This is why *collaris* is also popularly known as the sand badger.

There is only one species of hog badger but at least three races (groups with physical characteristics that vary slightly within a species) are found in most of mainland China, northeastern India, Burma, Sumatra, Thailand, and Vietnam. While common throughout most of its range, the hog badger is rarely

This drawing of a hog badger appeared in an encyclopedia published over one hundred years ago. It is an excellent likeness of collaris.

captured at night when it feeds. Nor is it easy to remove hog badgers from the deep fissures between boulders where they frequently spend the day.

Collaris' legs and tail are rather long for a badger. The average specimen weighs between fifteen and thirty pounds and measures about two feet. Pelts vary greatly in color. The fur may be bluish, grayish, or whitish, but the ears, face, throat, and long tail are always white.

Hog badgers are not keen sighted, nor is their sense of hearing acute. They employ their piglike snout as an antenna to detect danger and to locate food. While *collaris'* snout and color serve to distinguish it immediately from both *meles* and *taxus*, there are other obvious differences. For example, *meles'* tail is always the same color as the back fur and its claws are dark. As indicated, the hog badger's tail is always white while its claws are almost colorless.

Ferret Badgers

These are the most streamlined of all badgers. Their slender bodies are between thirteen and nineteen inches in length, and they rarely weigh more than three pounds.

There are three species of ferret badger. The Chinese ferret badger *(Melogale moschata)* is native to mainland China, Burma, Cambodia, Laos, Taiwan, and Vietnam. *Melogale orientalis*, the Javan ferret badger, is found in Borneo and Java. The Indian ferret badger, also known as the Burmese ferret badger, is distributed in parts of India, Burma, Nepal, Thailand, and Vietnam. Zoologists call this species *Melogale personata*.

All three species of ferret badger have long, pointed faces bearing masks. Unlike the Eurasian badger's facial mask, which is composed of stripes, that of the ferret badger is made up of black, whitish, or yellowish splotches. Other common characteristics of ferret badgers are an elongated gristly nose, naked at the tip, and a white or reddish stripe running down the back.

Javan and Indian ferret badgers are pale brown on the upperparts, the fur being slightly lighter below. This color pattern is unusual, other badgers being darker on their underparts than on the back and sides. The Chinese ferret badger wears the most colorful coat of the three species. Sometimes the fur on the stomach is orange shading to a dark brown on the back. Other individuals may have whitish to gray-brown hair on the back. Commercially, *moschata*'s fur is known as palini. It is used for collars and trimming.

Little study has been made of ferret badgers in the wild. However, it has been established that they are most numerous in lowlands. Field naturalists have also learned that the ferret badger is slow moving and tries to keep out of sight by "almost dragging itself on the ground." It has also been determined that all three species climb.

When seeking food, ferret badgers rely more on the nose than the ears and eyes. The trio is particularly fond of insects. While satisfying their appetites, ferret badgers consume thousands of insect pests. As a result, they are welcome in native huts. But if a visiting badger accidentally gets too close to a member of a household, the animal expresses its displeasure

Little is known about the ferret badgers, all of which bear close resemblance to each other.

by barking. If this warning is ignored, the badger launches a vicious attack on the human.

Malayan Stink Badger

Mydaus javanensis' popular name is most fitting. When excited, the Malayan stink badger (also called the teledu) raises its two- to three-inch fluffy tail and ejects a vile-smelling, pale-

No badger produces a more noxious fluid in its anal scent glands than the Malayan stink badger. Yet, strangely enough, the teledu's vile-smelling musk can be used to make sweet-smelling perfumes.

green fluid from its well-developed anal scent glands. *Javanensis* can douse an enemy sixteen feet away with this acrid liquid.

Dogs are often suffocated or blinded by the stink badger's musk. Yet, interestingly enough, the Javanese sultans of yesteryear employed the teledu's excretion to make women more alluring. After diluting the pungent liquid, the sultans compounded it with other ingredients to make perfume. Today, peoples throughout *javanensis*' range—Borneo, Java, Sumatra, and the North Natuna Islands—eat the Malayan stink badger's flesh. They remove the scent glands immediately after killing the animal.

Rarely more than twenty inches long, this species weighs from three to eight pounds. Except for a broad band of white (which may be incomplete) running down the long, mane-like fur onto

The skull of the teledu, the Malayan stink badger

the tail, *javanensis* is covered with blackish-brown fur. The face is pointed, the muzzle relatively long, the legs short and stout.

Not all Malayan stink badgers excavate setts. Many live with porcupines and share their dens. Actually, *javanensis* does not need to sublet. It has a special adaption for digging—the toes on the forefeet are joined by a web of muscles that extends to the base of the claws. However, even those individuals that dig setts do not make full use of the web. Malayan stink badger burrows are seldom more than two feet below the surface.

Palawan or Calamian Stink Badger

Practically nothing is known of the life history and habits of this resident of Palawan Island and the Calamian Islands Group in the Philippine Archipelago. However, zoologists have determined that the original classification of this species was incorrect. Formerly, *Suillotaxus marchei* was placed in the same genus (a group of related species) as the Malayan stink badger. But modern scientists, noting that *marchei* lacked the dorsal stripe of that species and also had different coloration, looked for other dissimilarities. They found them. For example, *marchei* is heavier toothed, shorter tailed, and smaller eared than *javanensis*.

Between a foot and a foot-and-a-half long, *marchei's* weight is approximately the same as that of the Malayan stink badger. Palawan badgers are brown to black on the upperparts with a sprinkling of white or silvery hairs on the back and head. Underparts are brown. The scent glands produce an extremely disagreeable fluid.

> "This is the point, to speak it flat and plain."
> —*Chaucer*

6 *Badger and Man*

IS THE badger a pest or a friend of man?

The answer to this question varies—it depends upon who is doing the answering. Sportsmen dislike the badger because it eats the eggs and young of ground-nesting game birds. On the other hand, agricultural experts consider the badger a most important factor in controlling rodents and destructive insects. However, no praise comes from farmers whose fields are raided or whose crops are flattened by badgers. As is to be expected, cowboys hate the animal that digs the holes in which horses stumble, throwing their riders. But most stockmen realize that, were it not for badgers, untold thousands of tons of vegetation would be eaten by rodents instead of being available to cattle, goats, and sheep.

According to a naturalist writing approximately one hundred years ago, badgers are ". . . bloodthirsty beasts and not only raid chicken yards and devour the inmates, but kill far more than they eat apparently for the sheer lust for bloodshed." Today, we know that, with the exception of rogue individuals, badgers do not attack domestic fowl. Nor is there any incontrovertible evidence that badgers kill lambs. The lambs that badgers are seen eating are the remains of predators' kills.

Gophers and ground squirrels are staple items in the diet of the American badger. Were it not for taxus, *those little vegetation-consuming rodents would cause much more damage to crops than they now do.*

Not only is the badger unjustly accused of destroying poultry and livestock but also it has had the misfortune to alienate former friends. When almond orchards were first established in California, badgers were welcome tenants in the groves because they preyed on nut-eating rodents. Originally, the grass between the rows of almond trees was cut by hand and it was a simple matter for the mowers to avoid the piles of dirt around the badger burrows. Now the groves are mowed by machines and the blades are often dulled, bent, or broken by the mounds. As a result, badgers are no more welcome in almond groves at the present time than they are in irrigation districts where flooding often is caused by their digging for rodents.

Meles and *taxus* should be thriving. To be sure, coats made of the long, relatively silky, silvery spring pelts of American badgers living in northern regions have become fashionable in recent years. On the other hand, few furriers still glue white-tipped badger hairs onto the pelts of other animals to manufacture imitation silver fox fur. Moreover, synthetic bristles, brushless shaving creams, and the electric razor have made badger-hair shaving brushes almost obsolete.

Not only is there a decrease in the demand for badger fur but also there is presently little call for badger meat. Salted and smoked badger hams are no longer considered the delicacy they once were throughout Europe. Incidentally, although certain Indian tribes held the badger sacred, as noted, it was eaten by other tribes. According to the journal kept by Osborne Russell, a trapper in the Old West during the 1830's, badgers "were much used by the Snake and Bannock Indians" for clothing and for food.

Despite the lack of interest in badger fur and meat today, the "eternal digger" is in danger. Fortunately, the American badger's situation is not as precarious as that of the Eurasian badger. Although *taxus* is shot, gassed, and trapped throughout its range, there is no concerted effort to exterminate the species. Nevertheless, many American badgers have died after eating poisoned carrion designed to lure and destroy sheep-killing coyotes. A large number of badgers living on the Great Plains have been gassed by devices intended to kill predators on livestock. Because *taxus* became the innocent victim of these deadly contraptions, certain types have been outlawed. Nevertheless, in some parts of its range, the American badger is subjected to unintentional persecution. A hardy, self-reliant creature, *taxus* can hold its own against its natural enemies (the golden eagle and the mountain lion) but it cannot cope with man's unreasoning use of traps and poison.

In England during the reign of Elizabeth I (1533–1603), a law was passed providing for the payment of bounties for badger heads. Finally, in 1973, an act protecting badgers was passed. But the badger was protected long before 1973. Many years

In order to keep rodents out of areas being replanted, British foresters fit special badger gates into forest fences to allow "Brock" a free passage. The smaller rodents lack the strength to open the gates.

ago, badger baiting was prohibited by law as an act of cruelty to animals. However, this ban was designed not to benefit the badger but rather the dogs that fought it.

When the Badger Act became law in 1973, no one paid any attention to a clause allowing the Ministry of Agriculture to grant licenses to kill badgers "for the prevention of disease." The clause would have remained unnoticed except for the discovery that *meles* was a carrier of bovine tuberculosis. This threatens Great Britain's cattle industry, while the spread of the disease among badgers is a major cause of concern to conservationists.

Originally, the Ministry engaged in the gassing of badger setts, but this procedure has given way to live trapping. Cages are set up outside setts and any animals captured are taken to laboratories, where specially trained veterinarians test them for bovine tuberculosis. Animals found to be infected are humanely destroyed.

While the gassing of setts was very unpopular with animal lovers and conservationists, the "trespassing" of Ministry badger-trapping teams has aroused resentment in rural England. Meanwhile, no explanation has revealed why outbreaks of bovine tuberculosis occur in some areas with high badger populations and not in others with a similar density. As the experts seek the answer to this puzzling question, groups of people concerned about the future of the Eurasian badger frequently remove the Ministry's traps from the entrance to setts.

At the present time, bovine tuberculosis has been eliminated from most of Great Britian. But the Ministry has not slackened its investigation into the part the badger plays in the spread of the disease.

Hopefully, science will find a way to insure the future of *meles*. Still, the chances are that badgers will be killed illegally by the misinformed in the future. This is most unfortunate. Like its New World kin, the Eurasian badger is one of man's best animal friends.

Index

Adventures in Apache Country, 19
Ainu, 13
Alabama, 17, 21
American badger
   appearance, 24, 26, 29
   courtship of, 47-48
   diet, 41, 43-44, 57
   fur, 27-29
   habitat, 39
   natural enemies, 59
   nurseries, 40-41
   persecution of, 59
   range, 10, 26
   species, 29
   weight, 26
Antarctica, 9
Arctic Circle, 25
Arctonyx collaris. See Hog badger
Asia, 10, 25
Audubon, John James, 22, 23
Australia, 9

Badger
   ancestry, 7, 9-10, 24
   in art, 22-23
   baiting, 18-19, 60
   bovine tuberculosis and, 60-61
   as charm, 14-15, 23
   emergence myths and, 17-18
   in literature, 21-22
   man and, 57-61
   meat of, 59
   medicinal uses of, 16-17, 43
   name derivation, 11
   as parents, 48-50
   place names and, 19-20, 21
   in popular speech, 19
   religion and, 18, 59
   species, 10
   strength, 32
   study of, 11
   superstitions concerning, 12-15, 23
   temperament, 45, 46
   weatherlore featuring, 15
Badger Clan, 18
Badger Protection Act, 59
"Badgers," 21
"Badger State," 21
Bears, 24
Berwick, Thomas, 22, 23
Bible, 21
Borneo, 16, 55
British Isles, 11, 26, 42, 61
"Brock," 11
Brock, Tommy, 22
Brown, John Ross, 19
Burma, 51, 53
Burmese ferret badger. *See* Indian ferret badger
Burrows, 31, 39, 40, 41, 56, 58

Calamian Islands, 56
Calamian stink badger. *See* Palawan stink badger
California, 17, 39, 58
Cambodia, 53
Canada, 22, 26
Carnivores, 7, 9, 26, 33

China, 51, 53
Chinese ferret badger
   appearance, 52, 53
   range, 53
Clare, Jean, 21
Claws, 28, 29, 30, 42, 45
Coyote, 45
Creeks, 17
Cumberland Cave, 26

Dachshund, 20
Death Valley, 39
Delayed implantation, 48
Denning, 46-47
Digging, 29-30, 31, 37-38, 40

"Earth stopping," 31
East Indies, 51
Eckert, Allan W., 22
Elizabeth I of England, 59
England, 17, 19, 59
Eurasian badger
   antiquity of, 10, 24
   appearance, 25-28
   courtship, 47-48
   diet, 42-43, 44, 57
   facial mask, 11, 28-29, 53
   habitat, 37
   housekeeping, 39
   range, 10, 24-25
   skin, 26
   species, 25
   strength, 32
   weight, 25-26
Europe, 10, 13, 15, 16, 21, 24, 33, 59

Far East, 51
Feet, 29-31
Ferret, 9
Ferret badgers, 52-54
Florida, 17
Fox, 15, 39
Frisch, Karl von, 42
Fur, 9, 14, 27-29, 43, 55-56, 58-59

Gait, 30
Georgia, 17
Germany, 15, 20
Grahame, Kenneth, 38

Graves, Robert, 21
Grison, 9
Grooming, 28, 30, 45-46
Groundhog Day, 15

Heck, Ludwig, 26
Hoban, Russell, 22
Hog badger
   appearance, 51-52
   range, 51
   senses, 52
Hopi, 17

Illinois, 20
Incident at Hawk's Hill, 22
India, 51, 53
Indian ferret badger
   appearance, 52-53
   range, 53
Indians, 17-18
Iowa, 20, 41
Ireland, 19
Island of Rhodes, 25
Israelites, 21
Italy, 14

Japan, 12, 13, 21, 25
Java, 16, 53
Javan ferret badger
   appearance, 52-53
   range, 53
Jaw, 33-34
Jicarillo Apache, 17

Lacrosse, 18
Lampe, Richard, 40
Laos, 53
Legs, 26, 29-31
Linnaeus, Carl, 24
Low Countries, 14

Madagascar, 9, 12-13
Madison *State-Journal*, 21
Malayan stink badger
   appearance, 54-56
   range, 55
Marten, 9
Maryland, 26
Meles meles. See Eurasian badger

Meles meles rhodius, 25
Melogale moschata. *See* Chinese ferret badger
Melogale orientalis. *See* Javan ferret badger
Melogale personata. *See* Indian ferret badger
Menomini, 18
Mexico, 10, 16, 26
Miacids, 7
Ministry of Agriculture, 60, 61
Mink, 9
Miocene, 10
Mohave, 17
Mustelidae, 9, 10, 51
Mustelids, 9-10, 44
Mydaus javanensis. *See* Malayan stink badger

Nepal, 53
Nevada, 40
New Forest, 26
New Mexico, 18
New Zealand, 9
North America, 10, 15
North Natuna Islands, 55

Oregon, 21
Otter, 9
Ovens, 39

Palawan Islands, 56
Palawan stink badger
  appearance, 56
  range, 56
Palini, 53
Philippine Archipelago, 56
Pima, 18
Pleistocene, 26
Polecat, 9
Potter, Beatrix, 21-22

Rabbits, 42
Ratel, 9
Rodents, 40, 44, 58
Russell, Osborne, 49

Sand badger. *See* Hog badger
Satan, 13, 14

Scent glands, 9, 35, 55, 56
Seals, 7
Senses
  hearing, 34, 35, 51, 53
  sight, 34, 52, 53
  smell, 34-35, 52, 53
Setts, 24, 31, 38-39, 40, 56
Shakespeare, 21
Shasta, 17
Skull, 33-34
Skunk, 9
Snakes, 41, 44
Somersetshire, 16-17
Sora, 9
Sporting Magazine, 16
Stink badgers, 52-54
Suillotaxus marchei. *See* Palawan stink badger
Sumatra, 21

Tail, 25, 26, 52
Taxidea taxus. *See* American badger
Tayra, 9
Teeth, 32-33, 45
Teledu. *See* Malayan stink badger
Texas, 19
Thailand, 51
Tracks, 30-31
Twain, Mark, 21

United States of America, 15, 19

Vesey-Fitzgerald, Brian, 36
Vietnam, 51, 53
Voice, 36, 54
Voles, 42

Walpole, Horace, 21
Warning coloration, 11
Weasel, 9
Wind in the Willows, The, 38
Wisconsin, 18, 20-21
Wolverine, 9
Woodchuck, 15, 44

Young, 35, 48-50

Zoril, 9
Zuni, 17, 18

WITHDRAWN